Science

WITHDRAWN

Assessment Book

Grade 2

Table of Contents

Practice Activities iii

Unit A — Plants and Animals
Chapter 1 Test—Plants 1
Chapter 2 Test—Animals 5
Unit A Performance Assessment 9

Unit B — Homes for Plants and Animals
Chapter 3 Test—Land Habitats 13
Chapter 4 Test—Water Habitats 17
Unit B Performance Asssessment 21

Unit C — Changes on Earth
Chapter 5 Test—Weather and Earth Changes 25
Chapter 6 Test—Earth Yesterday and Today 29
Unit C Performance Assessment 33

Unit D — The Sun and Its Family
Chapter 7 Test—The Sun and Earth 37
Chapter 8 Test—Moon, Stars, and Planets 41
Unit D Performance Assessment 45

Unit E — Matter and Energy
Chapter 9 Test—Matter 49
Chapter 10 Test—Energy 53
Unit E Performance Assessment 57

Unit F — Watch It Move
Chapter 11 Test—Force and Machines 61
Chapter 12 Test—Forces and Magnets 65
Unit F Performance Assessment 69

Scoring Chart 73

The McGraw-Hill Companies

Macmillan McGraw-Hill

Published by Macmillan/McGraw-Hill, of McGraw-Hill Education, a division of The McGraw-Hill Companies, Inc., Two Penn Plaza, New York, New York 10121.

Copyright © by Macmillan/McGraw-Hill. All rights reserved. The contents, or parts thereof, may be reproduced in print form for non-profit educational use with MacMillan/McGraw-Hill Science, provided such reproductions bear copyright notice, but may not be reproduced in any form for any other purpose without the prior written consent of The McGraw-Hill Companies, Inc., including, but not limited to, network storage or transmission, or broadcast for distance learning.

Printed in the United States of America
2 3 4 5 6 7 8 9 024 09 08 07 06 05 04

Name _____

Practice Activities
Units A–F

Practice Activities

Vocabulary

week   year   day

Use these words once for items 1–3.

1. There are 24 hours in a _____day_____.

2. There are seven days in a _____week_____.

3. There are 12 months in a _____year_____.

Circle the best answer.

4. For what do you use your ears?

    a. to taste

    b. to see

    (c.) to hear

5. Which picture shows an insect?

    a.    b.    (c.)

To the Teacher: Guide children through the activities above. Be sure children understand they need to use each vocabulary word once to complete activities 1–3. Introduce the multiple-choice questions by pointing out that answer options are labeled with different letters. Direct children to circle the letter that corresponds with the correct answer option. Point out that art may be used for answer options, as in activity 5.

Units A–F   Use before assessing.

Name _____

Mark true or false. If false, underline the word that makes the statement false.

6. __true__   Apples are a fruit that grows on trees.

7. __false__   Cats have two eyes and three legs.

8. __true__   Rulers are used to measure length.

9. Draw a line from the object to it's name.

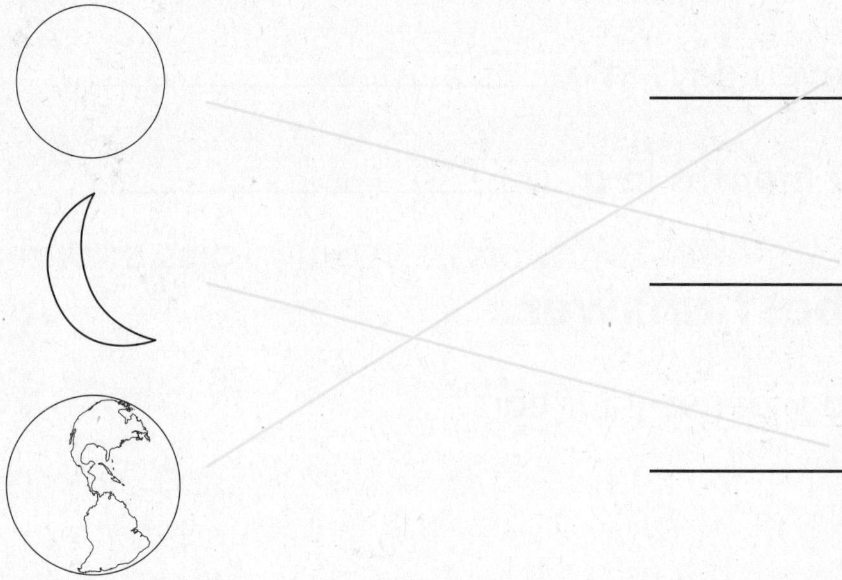

10. Draw a picture showing snowy weather. Draw a picture of rainy weather.   Children should draw one picture depicting snow and one depicting rain.

To the Teacher: Guide children through the activities.

Units A–F   Use before assessing.

Name _____

Chapter Test — Chapter 1

Plants

Vocabulary

pollen seeds oxygen
minerals life cycle

Use these words once for items 1–5.

1. Plant roots take in ____minerals____ from the soil.

2. New plants grow from ____seeds____.

3. You can see how a plant grows, lives, and dies by watching its ____life cycle____.

4. Plants help people breathe by making ____oxygen____.

5. Insects move from flower to flower carrying powdery ____pollen____.

Unit A · Plants and Animals — Use with textbook pages A2–A29

Name _____

Chapter Test
Chapter 1

6. Circle the picture that shows a living thing.

A    B    C⃝

Mark true or false. If false, underline the word that makes the statement false.

7. __false__ <u>Nonliving</u> things grow and change.

8. __true__ Nonliving things do not need food, water, and air.

9. __true__ Living things can make other living things like themselves.

10. __false__ Plants <u>cannot</u> make their own food.

Unit A · Plants and Animals   Use with textbook pages A2–A29

Name _____

Chapter Test — Chapter 1

Label the diagram with words from the box.

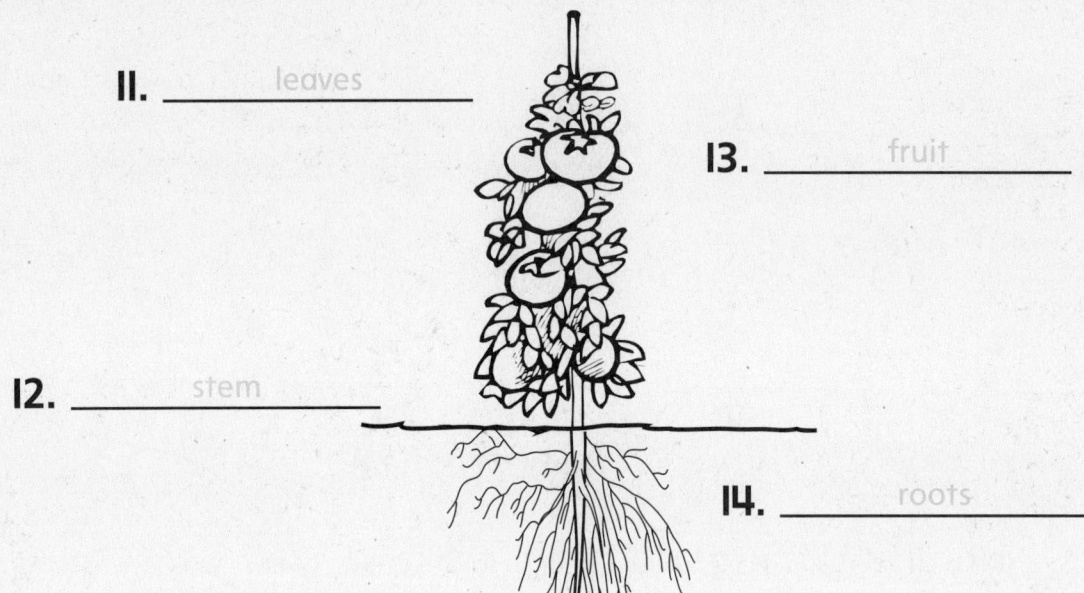

11. ___leaves___

12. ___stem___

13. ___fruit___

14. ___roots___

Vocabulary

roots   stem   leaves   fruit

15. Complete this sentence:

If there were no plants on the earth, people would not have as much

Accept any answer that children can prove with ideas from the chapter.
Sample answers: oxygen, food, medicine, clothing, building materials.

Unit A · Plants and Animals   Use with textbook pages A2–A29   3

Name _____

Chapter Test
Chapter 1

Color the picture that shows something that comes from a plant.

16. A 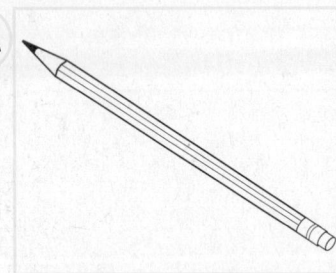 B 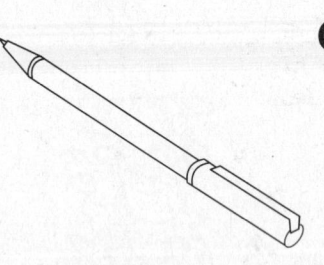 C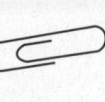

17. A  B  C

18. A  B  C

19. A 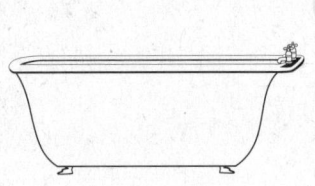 B 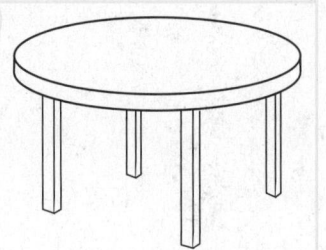 C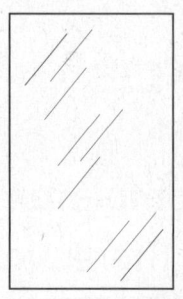

20. A  B  C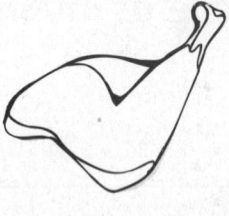

Unit A · Plants and Animals Use with textbook pages A2–A29

Name _____

Chapter Test — Chapter 2

Animals

Write the letter of the correct meaning on the line before each word.

__C__ 1. amphibian

__E__ 2. mammal

__A__ 3. predator

__B__ 4. prey

__D__ 5. larva

__F__ 6. insect

A an animal that hunts another animal

B an animal that is hunted by another animal

C an animal that can live both in water and on land

D an animal that is one stage in the life cycle of a butterfly

E an animal that has either hair or fur and breathes with lungs

F an animal that has three body parts and six legs

Unit A · Plants and Animals   Use with textbook pages A32–A53

Name _____

Chapter Test — Chapter 2

Look at the pictures. Complete the sentences.

7.  The fish is using its gills to get ____air____ from the water.

8. 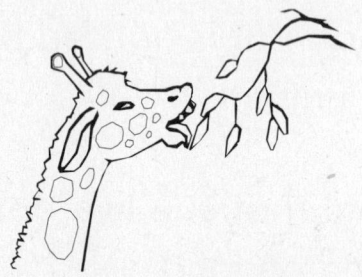 The giraffe eats leaves for ____food____.

9.  The horse is drinking ____water____.

10. 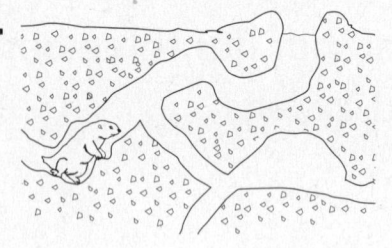 The prairie dog is in a place that is ____safe____.

Use sentences 7 through 10 to complete this sentence.

11. All animals need ____air____, ____food____, ____water____, and a ____safe____ place to live.

Unit A · Plants and Animals — Use with textbook pages A32–A53

Name _____

Chapter Test Chapter 2

Number the pictures to show how each animal changes.

12. A    B    C
    _3_   _1_   _2_

13. A    B    C
    _1_   _2_   _3_

14. A    B    C
    _3_   _2_   _1_

15. A  B  C
    _2_   _1_   _3_

16. A  B  C
    _3_   _2_   _1_

Unit A · Plants and Animals   Use with textbook pages A32–A53

Name _____

Chapter Test Chapter 2

Look at the pictures. Read the clues. Write the name of the animal group on the line.

17. ____Birds____ We have two legs. We have feathers.

18. ____Mammals____ We have hair or fur. We breathe with lungs.

19. ____Fish____ We breathe with gills. We use fins to swim.

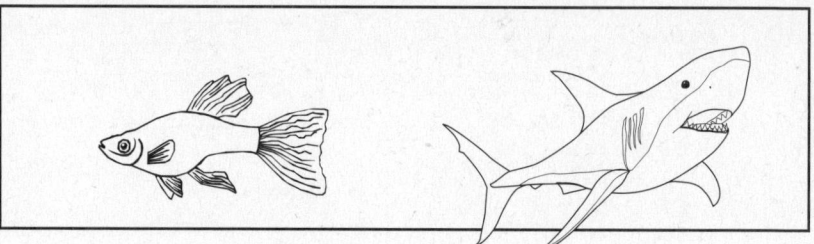

20. ____Reptiles____ Our skin is dry and scaly. Many of us lay eggs.

8 Unit A · Plants and Animals Use with textbook pages A32–A53

Teacher Resources for Unit Performance Assessment
Chapter 1

Dissect a Seed

Materials

- dried lima beans that have been soaked overnight in water, sharp knife, hand lens, pencils

Scoring Rubric

5 — **5 points** Student accurately draws and labels the seed's embryo (tiny plant), stored food, and seed coat. Student clearly explains the purpose of each seed part in the written answers. Student orally explains the parts of the drawing.

4 — **4 points** Student accurately draws and labels the seed's embryo, stored food, and seed coat. Student can orally explain the purpose of each seed part, but the written answers to the Analyze the Results questions are disorganized.

3 — **3 points** Student draws the seed parts sloppily but labels the parts accurately. Student writes partial answers, but orally points out the small plant in the seed and explains the purpose of at least one other seed part.

2 — **2 points** Student draws and labels some of the seed parts inaccurately. Student writes inaccurate answers to some of the questions, but can point out the small plant in the seed.

1 — **1 point** Student draws an unrecognizable drawing and does not write labels. Student fails to answer some of the questions but can point out the small plant in the seed.

0 — **0 points** Student does not complete the assignment.

Unit A · Plants and Animals | Use with textbook pages A2–A29

Compare Animals

Teacher Resources for Unit Performance Assessment
Chapter 2

Materials

- animal guides or picture books for several animal groups, pencil, crayons

Scoring Rubric

5 **5 points** Student chooses an animal group and two animals in the group to research. Student locates and reads the information accurately. Student draws animal pictures in the correct places and lists at least three relevant facts in each part of the diagram. Student can use the diagram to explain orally how the two animals are alike and different.

4 **4 points** Student chooses an animal group and two animals to research. Student locates and reads the information accurately. Student lists at least one fact in each part of the diagram. Student uses information from the diagram to compare the two animals orally.

3 **3 points** Student chooses an animal group and two animals to research. Student needs help locating information but writes at least one fact in each part of the diagram. Student can explain how the animals are alike and different.

2 **2 points** Student chooses an animal group and two animals to research, but needs help locating information. Student does not completely fill out each section of the diagram. Student can name the animals and their group, but cannot compare them orally.

1 **1 point** Student chooses animals from different groups to research. Student writes a few facts, but not in the correct parts of the diagram. Student is not able to explain that all animals in the same group share some characteristics but also have differences.

0 **0 points** Student does not complete the assignment.

Name _____

Unit Performance Assessment
Chapter 1

Dissect a Seed

What to do

Examine Watch as your teacher cuts apart the seed. Use the hand lens to examine the two halves carefully. Draw what you see. Label the tiny plant, the stored food, and the seed coat. Tell about the parts of your drawing.

What you need

dried lima beans that have been soaked overnight in water

hand lens

pencils

Analyze the Results

1. What is the job of the seed coat?

   The seed coat protects the tiny plant and its food.

2. What is the job of the stored food?

   The stored food gives the tiny plant the nutrition it needs until it gets big enough to make its own food.

3. What is the job of the tiny plant?

   The tiny plant grows into a new plant that will make new seeds.

Unit A · Plants and Animals   Use with textbook pages A2–A29

Name _____

Unit Performance Assessment
Chapter 2

Compare Animals

What to do

What you need

animal guides or picture books for several animal groups

pencil

crayons

Compare Choose two animals from the same group. Draw one of the animals over the first circle in the diagram. Draw the other over the second circle. Color the animals. Read books to find facts about each animal. Write facts that are true of only one of the animals in the circle below its picture. Write facts that are true of both animals in the overlapping part of the circles.

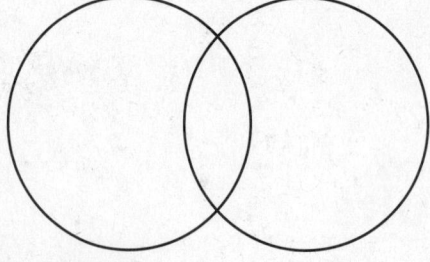

Analyze the Results

1. Which group do the animals you researched belong to?

   _____

2. In what ways are the two animals alike? Different?

   _____

   _____

   _____

Unit A · Plants and Animals   Use with textbook pages A32–A53

Name _____

Chapter Test
Chapter 3

Land Habitats

Vocabulary

Arctic   desert   migrate
rain forest   woodland forest

Read the clue. Write the word from the box that best completes each sentence.

1. It is a hot, dry habitat.
   It is a ____desert____.

2. It is a habitat that gets wet almost every day.
   It is a ____rain forest____.

3. It is a cold habitat.
   It is the ____Arctic____.

4. Some animals do this to escape a cold habitat in the winter.
   They ____migrate____.

5. It is a habitat that gets enough rain and light for trees to grow well. It is a ____woodland forest____.

Unit B · Homes for Plants and Animals   Use with textbook pages B2–B29   13

Name _____

Chapter Test
Chapter 3

6. Circle the letter of the answer that best completes this sentence:

   A habitat is

   A something people and animals do over and over.

   (B) a place where plants and animals live.

   C a covering that keeps plants and animals warm.

Write food or shelter to tell which need the animal in each picture is meeting in its habitat.

7.  ____food____

8.  ____shelter____

9.  ____shelter____

10.  ____food____

Unit B · Homes for Plants and Animals   Use with textbook pages B2–B29

Name _____

Chapter Test
Chapter 3

11. Color the animal that lives in a desert.

A  B  C

12. Color the animal that lives in the Arctic.

A  B  C

13. Color the animal that lives in a woodland forest.

A  B  C

14. Color the animal that lives in a rain forest.

A  B  C

Unit B · Homes for Plants and Animals · Use with textbook pages B2–B29

Name _____

Chapter Test Chapter 3

15. What do you think animals do when their habitat changes? Write one reason to support your answer.

Children's answers should indicate that the animals would die or move away since they will no longer be able to meet their needs.

Cross out the picture that does not belong in each habitat.

16. A    B    C

17. A    B    C

18. A    B    C

19. A    B    C

20. A 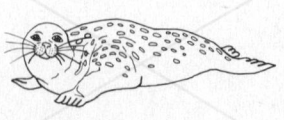   B    C

Unit B · Homes for Plants and Animals   Use with textbook pages B2–B29

Water Habitats

Write the letter of the correct meaning on the line before each word.

__E__ 1. ocean

__C__ 2. pollution

__A__ 3. pond

__B__ 4. recycle

__D__ 5. stream

A a body of fresh water that stays in one place

B waste that can be made into new things and used again

C waste that harms land, water, or air

D a moving body of fresh water

E a large body of salt water

Name _____

Chapter Test — Chapter 4

Vocabulary

| salt | food web | pollution |
| stream | insects | |

Write the word from the box that best completes each sentence.

6. Fish that live in a pond can eat water plants and ___insects___.

7. Salmon can move from place to place by swimming in a ___stream___.

8. Ocean animals need to live in water that has ___salt___.

9. Plankton are a part of the ocean ___food web___.

10. Water habitats can be destroyed by ___pollution___.

Unit B · Homes for Plants and Animals   Use with textbook pages B32–B53

Name _____

Chapter Test
Chapter 4

Color the picture in each pair that shows someone helping to care for the air, land, or water.

11. A    B

12. A     B

13. A    B

14. A 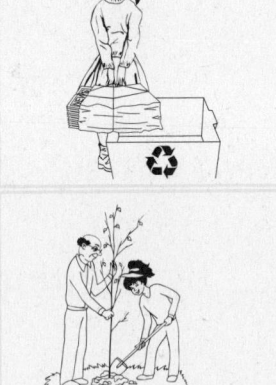   B

15. What do you think will happen if people do not stop pollution? Write one reason to support your answer.

Children's answers should indicate that many plants and animals would

die because their land, air, or water habitats would be spoiled.

Unit B · Homes for Plants and Animals — Use with textbook pages B32–B53

Name _____

Chapter Test — Chapter 4

Complete.

16. Oil spills are a form of _____pollution_____

17. We care for the _____water_____ by not dumping things in the ocean.

18. We care for the _____air_____ by riding bikes and walking.

19. We care for the _____land_____ by planting trees

20. What things can you recycle?

_____Possible answers: paper, glass, cans, plastic_____

Teacher Resources for Unit Performance Assessment

Chapter 3

Draw an Animal in Its Habitat

Materials

- 1 index card per child with the name of an animal, drawing paper, crayons

Scoring Rubric

5 **5 points** Student accurately draws and titles the habitat picture and fills the page with details that show different parts of the habitat. Student mentions ways for the animal to find food and shelter in the habitat.

4 **4 points** Student accurately draws the habitat without including details that belong in a different habitat. Student names one source of food for the animal and one source of shelter.

3 **3 points** Student names the correct habitat, but includes only a few details in the drawing of the habitat. Student fails to mention food or fails to mention shelter.

2 **2 points** Student names the correct habitat, but includes details from more than one habitat in the drawing. Student mentions a way to find food or a way to find shelter, but not both.

1 **1 point** Student draws the incorrect habitat for the animal, but does indicate logical ways for the animal to meet its needs for food and shelter.

0 **0 points** Student does not complete the assignment.

Observe Pollution

Teacher Resources for Unit Performance Assessment
Chapter 4

Materials

- plastic drinking glasses, water, food coloring, lemon juice, celery stalks with a fresh cut at the bottom

Scoring Rubric

5   **5 points** Student follows directions to perform the experiment and remembers to check the progress. Student keeps a log of changes. Student notes color and taste changes and gives a conclusion about how polluted water affects the animals and plants around it.

4   **4 points** Student follows directions to perform the experiment and keeps a log of changes. Student may need help making the connection between polluted water and its effect on animals that rely on the plants that use it.

3   **3 points** Student follows directions to perform the experiment. Student may occasionally forget to check the experiment's progress and record observations. Student notes that the plant looks and tastes different but may not make the leap from food coloring and lemon juice to real pollutants that may be poisonous.

2   **2 points** Student needs help to perform the experiment and fails to record observations in the log. Student answers the literal questions correctly, but does not attempt to answer question 3.

1   **1 point** Student leaves out the food coloring or the lemon juice while performing the experiment. As a result, the student is not able to answer the questions correctly.

0   **0 points** Student does not complete the assignment.

Name _____

Unit Performance Assessment
Chapter 3

Draw an Animal in Its Habitat

What to do

Communicate Take the card your teacher gives you. Draw a picture of the animal named on the card in the middle of a sheet of paper. Write a title for your picture that includes the name of the animal and its habitat. Draw and color details of the animal's habitat.

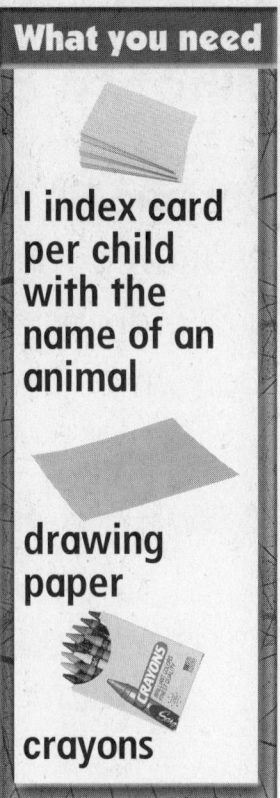

What you need

1 index card per child with the name of an animal

drawing paper

crayons

Analyze the Results

1. What animal did you draw?

   _____

2. What habitat did you draw?

   _____

3. How does the animal meet its needs in the habitat?

   Children's answers should include at least one way animals can find food in
   _____
   the habitat and one way animals find shelter.

Unit B · Homes for Plants and Animals   Use with textbook pages B2–B29   23

Name _____

Unit Performance Assessment

Chapter 4

Observe Pollution

What to do

Observe Fill the glass with clear water. "Pollute" the water with food coloring and a few drops of lemon juice. Then stand up a stalk of celery in the glass. Check on the celery every half hour. Keep a log of the changes.

What you need

Plastic cups

Food coloring

Celery

Analyze the Results

1. How did the "pollution" change the way the plant looks?

   It changed the plant's color.

2. How did the "pollution" change the plant's taste?

   It made the celery taste lemony.

3. How do you think dangerous pollution would affect animals who eat plants that live near polluted water?

   It might poison the animals or make them stop eating good food because it tastes different.

Name _____

Chapter Test — Chapter 5

Weather and Other Earth Changes

Vocabulary

| condenses | earthquake | erosion | evaporate |
| landslide | precipitation | volcano | water vapor |

Use these words once for items 1 to 8.

1. Snow, rain, and hail are kinds of __precipitation__.

2. The hot sun can make water __evaporate__.

3. Wind breaking down rock is called __erosion__.

4. When plates in the Earth's crust move, it can cause an __earthquake__.

5. A mountain made of cooled lava is a __volcano__.

6. The quick movement of soil down a hill is a __landslide__.

7. When water vapor changes to a liquid, it __condenses__.

8. Bubbles from boiling water go into the air as __water vapor__.

Unit C · Changes on Earth — Use with textbook pages C2–C27

Name _____

Chapter Test
Chapter 5

9. Draw a line from the weather picture to its name.

Tornado   Volcano   Hurricane

10. Draw a line under heat sources that can cause water to evaporate.

A    B 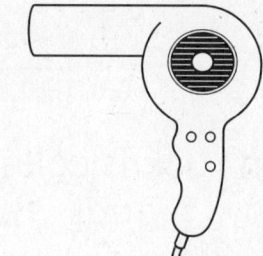   C

Circle the best answer.

11. What causes a drought?
    A not enough sun   C too much rain
    (B) not enough rain   D too much snow

12. What can both tornadoes and earthquakes cause?
    F erosion   (H) landslides
    G fast winds   J hurricanes

Unit C · Changes on Earth   Use with textbook pages C2–C27

13. What is lava?

   A ash
   (B) melted rock that hardens when it cools
   C moving plates that cause an earthquake
   D hot gases

14. What do you call water that evaporates into the air?

   F the water cycle
   G precipitation
   H ice
   (J) water vapor

15. What is the first step in the water cycle?

   A Water vapor rises.
   (B) Heat from the Sun evaporates water.
   C Precipitation falls.
   D Drops of water form clouds.

16. What helps prevent erosion of soil?

   (F) plants
   G sand
   H wind
   J rain

17. What can cause erosion of a sandy beach?

   A beach dumping
   B sea life
   (C) ocean waves
   D the Sun in summer

Unit C · Changes on Earth — Use with textbook pages C2–C27

Name _____

Chapter Test Chapter 5

Look at the diagram of the water cycle.

Complete each step in the water cycle.

18. Heat from the Sun evaporates ____water____.

19. Water vapor ____rises____. It cools and condenses into tiny drops of water.

20. The drops of water form ____clouds____.

21. When the drops of water get big enough, they fall to the ground as ____precipitation____.

22. Water returns to the lakes and ____oceans____. The cycle begins again.

28 Unit C · Changes on Earth Use with textbook pages C2–C27

Name _____

Chapter Test
Chapter 6

Earth Yesterday and Today

Vocabulary

endangered   extinct   fossils
paleontologist   skeleton

Use these words once for items 1 to 5.

1. A scientist who studies dinosaurs is a ___paleontologist___.

2. Remains of living things from a long time ago are called ___fossils___.

3. Living things that are close to becoming extinct are ___endangered___.

4. A full set of dinosaur bones can be put together into a ___skeleton___.

5. Living things no longer on Earth are ___extinct___.

Circle the best answer.

6. Which picture shows a dinosaur skeleton?

A    B 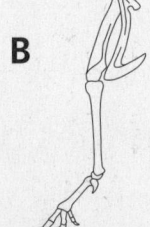   (C)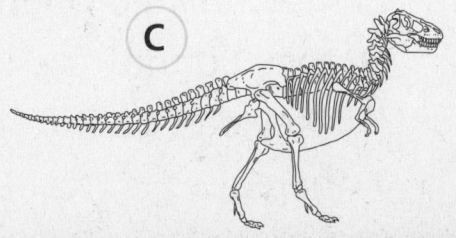

Unit C · Changes on Earth   Use with textbook pages C28–C53

Name _____

Chapter Test
Chapter 6

7. Circle the picture that shows the first step in a fossil being formed.

A    B    C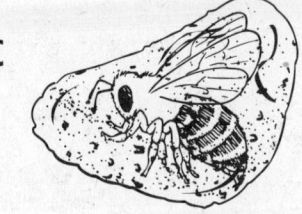

(A is circled)

8. Underline the animal that was saved from being extinct.

A    C

B    D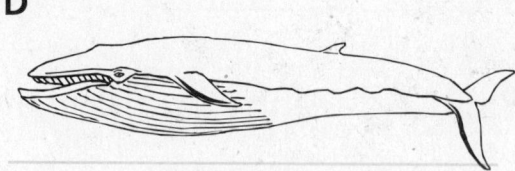

Circle the best answer.

9. What do we call a 3-million year old tooth found in the dirt?
   A an animal eater   C a dinosaur
   B endangered   (D) a fossil

10. What is the same about dinosaurs and animals that live today?
    F they only eat plants
    G they cannot live in the water
    (H) they can be hurt by changes on Earth
    J they always travel in groups

30   Unit C · Changes on Earth   Use with textbook pages C28–C53

Name _____

Chapter Test
Chapter 6

Circle the best answer.

11. What can a paleontologist learn from a dinosaur skeleton?
    - (A) the size of the dinosaur
    - B what year it lived
    - C the color of the dinosaur
    - D how many babies it had

12. How can we protect endangered animals?
    - F make them extinct
    - (G) keep them in safe places
    - H hunt them
    - J make them house pets

13. Why do scientists compare fossil bones to bones of living animals?
    - (A) to figure out how the animals lived
    - B to figure out the name of the animal
    - C to figure out which animal is bigger
    - D to figure how which animal died first

14. What does long, sharp teeth tell about a dinosaur?
    - F It was big in size.
    - G It laid eggs.
    - H It had long legs.
    - (J) It ate meat.

Name _____

Chapter Test
Chapter 6

15. Which fact about dinosaurs is true?
    A All could fly.
    (B) Many lived in the water
    C None ate plants.
    D All laid eggs.

16. Why are fossils helpful to scientists?
    F Fossils are easy to find.
    (G) Fossils give clues about dinosaurs.
    H Fossils give clues to the future.
    J Fossils are always of complete skeletons.

17. Which object can become a fossil?
    (A) a tooth   C a roller blade
    B a hat   D a letter

18. Where did some dinosaurs live?
    F in an igloo   (H) in the sea
    G in a hut   J in the zoo

32   Unit C · Changes on Earth   Use with textbook pages C30–C53

Picture a Volcano

Teacher Resources for Unit Performance Assessment — Chapter 5

Materials

- large white drawing paper, markers, pencils, ruler

Scoring Rubric

5 — **5 points** Student illustrates a cone-shaped volcano that is erupting. Three or more parts are labeled, such as lava, gases, and ash. Student mentions that lava comes out of the volcano and becomes hard rock when cool. One other change is mentioned, such as resulting floods or landslides or new mountains created from volcanoes.

4 — **4 points** Student illustrates a cone-shaped volcano that is erupting. Two or more parts of the volcano are labeled. The student answers are partly correct, but flawed.

3 — **3 points** Student draws an erupting volcano but does not label more than one part. Answers to questions show some knowledge of what occurs during a volcano eruption, but are illogical.

2 — **2 points** Student draws a volcano but cannot label any of its parts. Response to questions shows no understanding of how the earth changes from a volcano.

1 — **I point** Student draws a volcano but cannot label any of its parts. Student does not answer the questions.

0 — **0 points** Student does not complete the assignment.

Save the Animal Poster

Teacher Resources for Unit Performance Assessment — Chapter 6

Materials

- books or other sources on endangered animals, poster paper, markers

Scoring Rubric

5 **5 points** Student makes a poster that gives facts about the animal's habitat and tells why it is endangered. Student explains the cause of the animal's endangered status and uses sound reasoning to suggest why it might or might not become extinct.

4 **4 points** Student makes a poster that gives facts about the animal's habitat and why it is endangered. Student gives an explanation for the status that is partly correct, but flawed. The student gives an opinion on the animal's possible extinction, but does not explain his or her reasoning.

3 **3 points** Student makes a poster that encourages people to save the animal, but it is not supported by accurate facts. The responses to the questions show some knowledge, but are flawed.

2 **2 points** Student makes a poster that is not complete or accurate. The responses to the questions are illogical.

1 **1 point** Student makes a poster that is not complete or accurate. The student does not answer the questions.

0 **0 points** Student does not complete the assignment.

Name _____

Unit Performance Assessment

Chapter 5

Picture a Volcano

What happens when a volcano erupts?

What to do

Communicate Draw a picture of a volcano that is erupting. Show what comes out of the volcano. Label each part in your drawing.

What you need

large white drawing paper

markers

pencils

ruler

Analyze the Results

How can a volcano change Earth?
List changes that can happen.

An erupting volcano can kill plant and animal life near the volcano. The lava that comes out of the volcano can harden into rock. Plant life and wildlife return to the area years later.

_____

_____

_____

Unit C · Changes on Earth   Use with textbook pages C2–C27

Name _____

Unit Performance Assessment

Chapter 6

Save the Animal Poster

Choose an endangered animal to learn about. Find out why the animal is endangered. Think about how the animal could be saved.

What you need

books or other sources on endangered animals

poster paper

markers

What to do

Communicate Make a poster that tells how to save the animal. Tell why it is in danger. Give facts about the animal's habitat.

Analyze the Results

What makes an animal endangered? Do you think your animal will become extinct? Tell why or why not.

Possible answers include: They are not able to meet their needs because their habitats are being destroyed; people also hunted them. Accept reasonable responses for the second part of the question.

36   Unit C · Changes on Earth   Use with textbook pages C28–C53

Name _____

Chapter Test
Chapter 7

The Sun and Earth

Vocabulary

axis   equator   orbit
rotate   Sun

Use these words once for items 1 to 5.

1. The star that gives Earth light and heat is the ____Sun____.

2. The path Earth takes as it moves around the Sun is its ____orbit____.

3. The imaginary line that separates the northern part of Earth from the southern part is the ____equator____.

4. It takes 24 hours for Earth to spin around, or ____rotate____.

5. Earth rotates on an imaginary line called an ____axis____.

Circle the best answer.

6. What happens when Earth spins on its axis?
   A) It rotates.
   B It orbits.
   C It moves toward the Sun.
   D It does not move.

Unit D · The Sun and Its Family   Use with textbook pages D2–D21   37

Name _____

Chapter Test
Chapter 7

7. Which star is closest to the Earth?
   F the Big Dipper
   G Venus
   (H) the Sun
   J the Moon

8. What would it be like on Earth without the Sun?
   A hot   (C) cold
   B warm   D light

9. Why does the Sun look small?
   F It is smaller than Earth.
   (G) It is so far away.
   H It is so hot.
   J It is smaller than the Moon.

10. How long does it take Earth to rotate once?
    (A) 24 hours
    B 7 days
    C 3 months
    D 365 days

11. Why is it dark here when it is light on the other side of Earth?
    F because the Sun is not always shining
    (G) because the Sun lights half of Earth at a time
    H because the Moon blocks the Sun
    J because the Sun is moving

38 Unit D · The Sun and Its Family   Use with textbook pages D2–D21

Name _____

Chapter Test
Chapter 7

12. Why does one part of Earth get more light and heat in the summer?
    - (A) that part is tilting toward the Sun
    - B that part is tilting away from the Sun
    - C the Sun is lower in the sky
    - D the Sun is bigger

13. What makes day and night?
    - (F) the rotating of Earth
    - G the seasons
    - H the hands on clocks
    - J Earth's orbit around the Sun

14. When it is summer north of the equator, which season is it south of the equator?
    - A summer
    - B fall
    - (C) winter
    - D spring

15. How long does it take Earth to orbit the Sun?
    - F 24 hours
    - G one season
    - H 300 days
    - (J) one year

16. Which places have the same weather most of the year?
    - (A) Places near the equator.
    - B Places north of the equator.
    - C Places south of the equator.
    - D Places near the ocean.

Unit D · The Sun and Its Family   Use with textbook pages D2–D21

Name _____

Chapter Test — Chapter 7

17. What is Earth's axis?
    F the way Earth orbits the Sun
    (G) an imaginary line from the North Pole to the South Pole
    H an imaginary line near the equator
    J an imaginary line north of the equator

Look at the Sun as it appears at different times of the day.

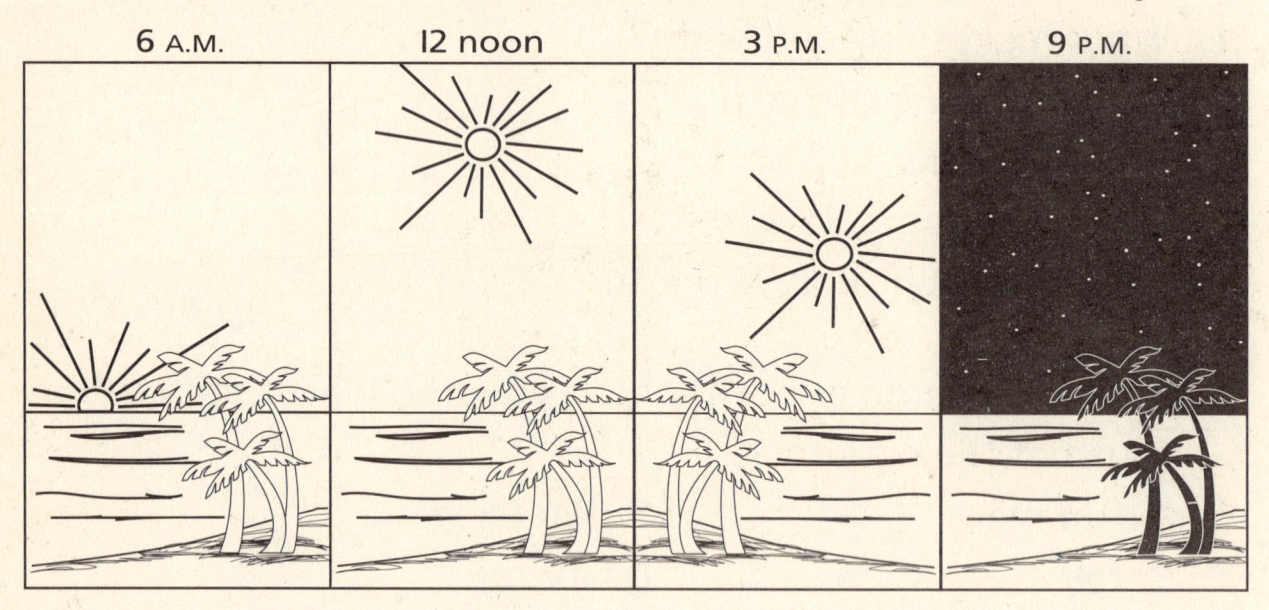

6 A.M.   12 noon   3 P.M.   9 P.M.

Use the picture. Complete.

18. At what time is the Sun giving the most light? __noon__

19. People on the beach cannot see the Sun at 9 P.M. because Earth has __turned away from the Sun__.

20. Is the beach getting more light at 6 A.M. or at 3 P.M.?
    __3 P.M.__

21. The Sun seems to move in the sky because Earth is __rotating__.

40 Unit D · The Sun and Its Family   Use with textbook pages D2–D21

Name _____

Chapter Test
Chapter 8

Moon, Stars, and Planets

Vocabulary

| constellation | craters | Moon | phases |
| planet | solar system | star | |

Use these words once for items 1 to 7.

1. The different shapes that the Moon seems to have as it orbits Earth are called ____phases____.

2. The Sun, nine planets, and their moons make up the ____solar system____.

3. A star pattern that makes a picture is a ____constellation____.

4. The name we give to a large object that orbits around the Sun is a ____planet____.

5. The ball of rock that orbits Earth is the ____Moon____.

6. The large holes in the Moon made by space rocks are ____craters____.

7. A hot ball of light that appears tiny in the night sky is a ____star____.

Unit D · The Sun and Its Family   Use with textbook pages D22–D53   41

Name _____

Chapter Test
Chapter 8

Complete.

8. The Moon's surface is made of mountains, smooth areas, and ____craters____.

9. How many planets are in the Solar System? __9__

Circle the best answer.

10. What is the brightest object in the night sky?
    A The Sun   C The Little Dipper
    (B) The Moon   D Earth

11. About how long does it take the Moon to change from a new moon to a full moon and back again?
    F one week   H 14 days
    (G) one month   J one year

12. When can we see a full moon?
    A during the first phase of the Moon
    B when the Moon is between the Sun and Earth
    C when the Moon's orbit is complete
    (D) when Earth is between the Moon and the Sun

13. How does the Moon get light?
    F It makes its own light.
    G Earth's light bounces off it.
    (H) The Sun's light bounces off it.
    J From the craters on it.

Name _____

Chapter Test
Chapter 8

Circle the best answer.

14. What tool helps scientists look at the night sky?
    - (A) a telescope
    - B a big dipper
    - C a star finder
    - D a constellation

15. What are some constellations named after?
    - F planets
    - (G) animals
    - H the Sun
    - J oceans

16. What can make a star in the sky look very bright?
    - A Being red in color.
    - (B) Being close to Earth.
    - C Being cold.
    - D Being small in size.

17. Which planet takes the most time to orbit the Sun?
    - F The Moon.
    - (G) The planet furthest from the Sun.
    - H The planet closest to the Sun.
    - J A planet with more than one moon.

18. Which is the largest planet in our solar system?
    - A Pluto
    - B Earth
    - C Venus
    - (D) Jupiter

Unit D · The Sun and Its Family   Use with textbook pages D22–D53   43

Name _____

Chapter Test
Chapter 8

19. Why do scientists use space probes?

   (F) To explore other planets

   G To learn about the seasons

   H To explore the ocean floors

   J To learn about how dinosaurs lived

20. Look at these pictures of the Moon. The new Moon is labeled I. It is the first phase. Which view of the Moon would you see next? Write the numbers 2, 3, and 4 to put the views of the Moon in order.

   _____ 3
   _____ 2
   _____ 4
   _____ 1

44 Unit D · The Sun and Its Family   Use with textbook pages D22–D53

Teacher Resources for Unit Performance Assessment
Chapter 7

Earth Needs the Sun

Materials

- white paper for writing and illustrating, construction paper for making a cover

Scoring Rubric

5 — **5 points** Student creates a book that tells why the Sun is important to Earth. Both the Earth's orbit around the Sun and the Earth's rotation are included. Illustrations show a relationship between Earth and the Sun. Answers to questions show a solid understanding of the Sun's importance.

4 — **4 points** Student creates a book about the Sun. The text is accurate but partly incomplete. Illustrations of Earth and the Sun do not show a relationship. Answers to questions show an understanding of the Sun's importance.

3 — **3 points** Student creates a book about the Sun with little mention of Earth. Illustrations are incomplete. Answers to questions show some understanding, but the answers are flawed.

2 — **2 points** Student creates a book about the Sun with little or no mention of Earth. Illustrations are inaccurate. The student's answers to questions are flawed.

1 — **1 point** Student creates a book about the Sun without any factual drawings. Descriptions are incorrect. The student does not complete the questions.

0 — **0 points** Student does not complete the assignment.

Unit D · The Sun and Its Family   Use with textbook pages D2–D21

Take a Trip to Space!

Teacher Resources for Unit Performance Assessment — Chapter 8

Materials

- large drawing paper, markers, old magazine or books with pictures of space

Scoring Rubric

5 — **5 points** Student makes a flyer that features a logical journey. Description shows a solid grasp of chapter vocabulary and content. Student uses facts about the solar system to explain the feasibility of the trip.

4 — **4 points** Student makes a flyer that features a logical journey. Description shows some knowledge of chapter vocabulary and content. Student uses facts about the solar system to explain the feasibility of the trip.

3 — **3 points** Student makes a flyer that features a logical journey. Description does not show a grasp of chapter vocabulary and content. The answer to the question includes facts, but is flawed.

2 — **2 points** Student makes a flyer describing a journey that does not show any grasp of chapter vocabulary and content. The response to the question is illogical.

1 — **1 point** Student makes a flyer describing a journey that does not show any grasp of chapter vocabulary and content. The student does not answer the question.

0 — **0 points** Student does not complete the assignment.

---

46 — Unit D · The Sun and Its Family — Use with textbook pages D22–D53

Name_____

Unit Performance Assessment
Chapter 7

Earth Needs the Sun

What to do

What you need

white paper

construction paper

Make a Model Make a book that tells why we need the Sun. Put the title "Earth Needs the Sun" on the cover. Explain how the Sun gives us daylight and the seasons. Draw pictures of Earth and the Sun to show how these things happen. Write facts that tell about your pictures.

Analyze the Results

1. What are the most important things we get from the Sun?

   The Sun gives Earth light and heat.

   _____

2. What would Earth be like without the Sun?

   Possible answers include: Cold and dark; there would be no plants
   or animal life on Earth; no day and night.

   _____

   _____

Unit D · The Sun and Its Family — Use with textbook pages D2–D21 — 47

Name _____

Unit Performance Assessment
Chapter 8

Take a Trip to Space!

What to do

Communicate Pretend you run trips to space. Decide where your next trip will be. You could land on the Moon or tour the solar system.

Prepare a flyer that describes the trip. Use words and ideas from the chapter to tell exactly where you will go and what you will see. Make the flyer exciting so people want to join you!

What you need

large drawing paper

markers

old magazine or books with pictures of space

Analyze the Results

Could scientists really make this trip and see and do what you described? Why or why not?

Accept all reasonable answers.

_____
_____
_____
_____
_____

48 Unit D · The Sun and Its Family   Use with textbook pages D22–D53

Name _____

Chapter Test
Chapter 9

Matter

Vocabulary

gas liquid mass matter
property solid temperature volume

Use these words once for items 1 to 8.

1. The color of an object is its _____property_____.

2. Anything that takes up space and has mass is _____matter_____.

3. Milk, a form of matter that takes the shape of its container, is called a _____liquid_____.

4. Using a cup to measure matter is one way to find its _____volume_____.

5. Air, a form of matter that spreads out to fill its container, is called a _____gas_____.

6. The amount of matter in an object is called _____mass_____.

7. A crayon, a form of matter that has its own shape, is called a _____solid_____.

8. When you measure how warm an object is, you name the object's _____temperature_____.

Unit E · Matter and Energy   Use with textbook pages E2–E27

Name _____

Chapter Test
Chapter 9

Circle the letter of the best answer.

9. Which is not a solid?

   A [cloud]   B [yo-yo]

   C [crayon]   D [fish]

10. Which is filled with a gas?

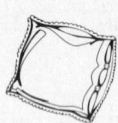

    F [teddy bear]   G [pillow/bag]

    H [balloon]   J [trash can]

11. Which is filled with a liquid?

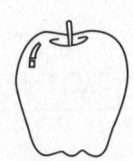

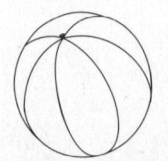

    A [tub]   C [apple]

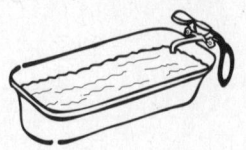

    B [ball]   D [toothpaste tube]

50 Unit E · Matter and Energy Use with textbook pages E2–E27

Name _____

Chapter Test
Chapter 9

Circle the type of change each picture shows.

12. physical change — (chemical change)

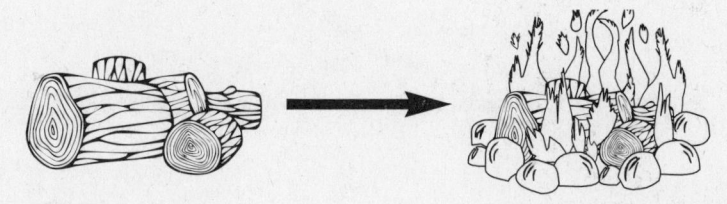

13. (physical change) — chemical change

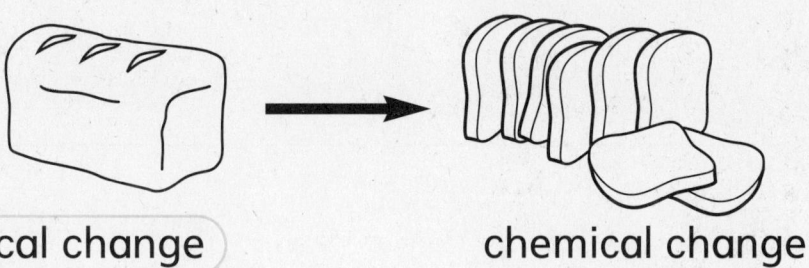

14. (physical change) — chemical change

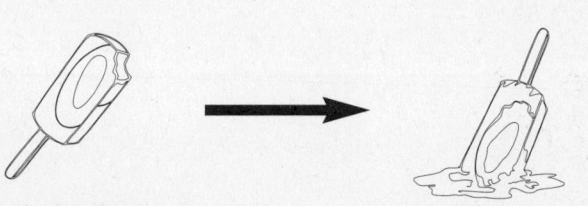

15. physical change — (chemical change)

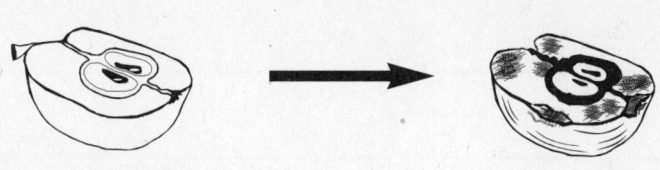

Unit E · Matter and Energy   Use with textbook pages E2–E27   51

Name _____

Chapter Test
Chapter 9

Name a solid, a liquid, and a gas in the picture. State a property of each kind of matter.

17. Solid

Answer may include tree, grass, flower, leaves.

Property

Answer will depend upon solid selected but may include that it has mass and a definite shape.

18. Liquid

Answer may include raindrops or puddle.

Property

Answer will depend upon liquid selected but may include clear or irregular shape.

19. Gas

Answer may include wind or air.

Property

Answer will depend upon gas selected but may include clear and has mass.

Unit E · Matter and Energy   Use with textbook pages E2–E27

Name _____

Chapter Test
Chapter 10

Energy

Vocabulary

energy   fuel   heat   light
pitch   reflects sound   vibrate

Use these words once for items 1 to 8.

1. Energy that you can hear is ____sound____.

2. Heat is one kind of ____energy____.

3. When light hits an object, some of it bounces off or ____reflects____.

4. To make heat, a fire needs to burn ____fuel____.

5. A musical instrument makes sound when its parts ____vibrate____.

6. The highness or lowness of a sound is its ____pitch____.

7. You can make solid ice cream melt by adding ____heat____.

8. A kind of energy that moves in straight lines is called ____light____.

Unit E · Matter and Energy | Use with textbook pages E28–E53 | 53

Name _____

Chapter Test
Chapter 10

Circle the letter of the best answer.

9. Which is not a kind of energy?
   A light
   (B) matter
   C heat
   D sound

10. Where does much of Earth's heat and light come from?
    F solids
    G air
    H fire
    (J) Sun

11. What happens when light moves through water?
    (A) Refraction happens.
    B A shadow forms.
    C The water becomes a solid.
    D The water becomes a gas.

12. What kind of sound do big vibrations make?
    F soft sounds
    G sounds with a low pitch
    (H) loud sounds
    J echo

Chapter Test
Chapter 10

Name _____

Write the answer on the line.

13. Name two ways that you use heat.

   Possible answers: to keep warm, to cook food, to change a state of matter

14. Name two sources of light.

   Possible answers: Sun, fire, electric lights

15. How is sound made?

   Sound is made when something vibrates.

Unit E · Matter and Energy — Use with textbook pages E28–E53

Name _____

Chapter Test
Chapter 10

Describe how heat is changing the state of water in each picture.

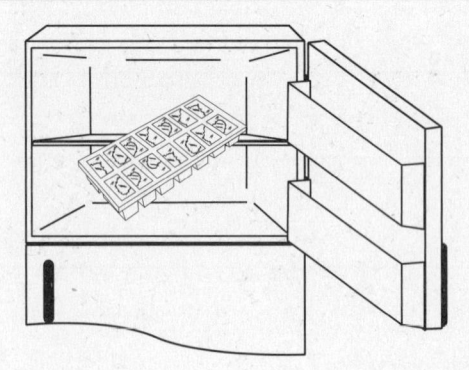

16.

_____

_____

17.

_____

_____

18.

_____

_____

56   Unit E · Matter and Energy   Use with textbook pages E28–E53

Most and Least Mass

Materials

- balance, small paper clip, quarter, box of chalk, notebook

Objects can vary slightly from the list given.

Scoring Rubric

5   **5 points** Student correctly measures the masses of three pairs of objects and records the results in the table. Student uses the data to correctly order the objects from least to most mass. Student clearly explains how his or her observations were used to form predictions.

4   **4 points** Student correctly measures the masses of two pairs of objects and records the results in the table. Student uses the data to correctly order the objects from least to most mass. Student clearly explains how his or her observations were used to form predictions.

3   **3 points** Student correctly measures the masses of two pairs of objects and records the results in the table. Student uses the data to correctly order the objects from least to most mass. Student does not explain how his or her observations were used to form predictions.

2   **2 points** Student makes one error while measuring. This causes a subsequent error in ordering the objects from least to most mass. Answers to questions are illogical or do not answer the questions.

1   **1 point** Student measurements are flawed and does not correctly order the objects from least to most mass. Student does not answer the questions.

0   **0 points** Student does not complete the assignment.

Observe Sounds

Teacher Resources for Unit Performance Assessment
Chapter 10

Materials

- assortment of unbreakable glass jars, wooden spoon, water

Objects can vary slightly from the list given.

Scoring Rubric

5   **5 points** Student lists observations of the sounds made by varying the water levels in the jars. Student clearly explains his or her answers to the questions. Student infers water level affects vibration rate, which determines pitch of the sound produced.

4   **4 points** Student lists observations of the sounds made by varying the water levels in the jars. Student clearly explains his or her answers to the questions. Student response to the inference is flawed but partially correct.

3   **3 points** Student lists observations of the sounds made by varying the water levels in the jars. Student response to the questions and the inference is flawed but partially correct.

2   **2 points** Student lists observations of the sounds made by two different water levels in the jars. Answers to questions and inference are illogical or do not answer the questions.

1   **1 point** Student lists observations of the sounds made by only one water level in the jars. Answers to questions and inference are illogical or do not answer the questions.

0   **0 points** Student does not complete the assignment.

Name _____

Unit Performance Assessment

Chapter 9

Most and Least Mass

What to do

Predict Your teacher has given you several objects. Make a prediction. List the objects in order from least mass to most mass. Record your predictions.

Prediction: _paper clip, quarter, box of chalk, notebook_

Compare Use the balance to compare two objects. Then compare another pair. Compare a third pair. Record the results.

What you need
- balance
- small paper clip
- quarter
- box of chalk
- notebook

Objects Being Compared	Object with Less Mass

Analyze the Results

1. Use the table to order the objects from least mass to most mass.

   paper clip, quarter, box of chalk, notebook

2. Compare your predictions with your measurements. Did you predict the correct order?

   Accept reasonable answers

Unit E · Matter and Energy — Use with textbook pages E2–E27

Observe Sounds

What to do

Observe Fill the first jar half way with water. Gently tap the side of the jar with the spoon. Observe the sound it makes. Add water to the jar so it is almost full. Tap it again. Observe the sound. Then pour out most of the water so that only a little remains. Tap the jar again. Observe the sound. Do again with the second jar. Record your observations in the table.

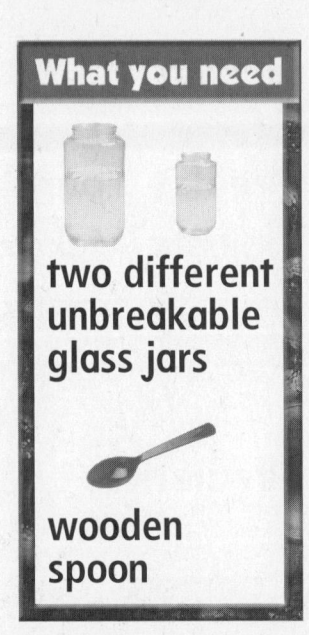

What you need

two different unbreakable glass jars

wooden spoon

	Jar 1	Jar 2
Sound made when half full		
Sound made when full		
Sound made when almost empty		

Analyze the Results

1. Which amount of water made a sound with the lowest pitch?

   Students will likely describe the jar with the most water.

2. Which amount of water made a sound with the highest pitch?

   Students will likely describe the jar with the least water.

Name _____

**Chapter Test
Chapter 11**

Forces and Machines

Vocabulary

| force | friction | fulcrum | gravity |
| lever | ramp | simple machine |

Use these words once for items 1 to 7.

1. A pair of scissors is a kind of simple machine called a _____lever_____.

2. When two moving things rub together, they slow down due to _____friction_____.

3. A simple machine with a slanted surface is a _____ramp_____.

4. It is easy to move a load from one place to another with a ___simple machine___.

5. A push or a pull is called a _____force_____.

6. A lever is a bar that rests on a fixed point or _____fulcrum_____.

7. Raindrops fall toward Earth due to _____gravity_____.

Unit F · Watch It Move — Use with textbook pages F2–F33 — 61

Name _____

Chapter Test
Chapter 11

Circle the letter of the best answer.

8. How are a push and gravity alike?
   - A Both are directions.
   - B Both are matter.
   - C Both are motions.
   - (D) Both are forces.

9. Which object needs the greatest force to move?
   - F A leaf
   - (G) A bicycle
   - H A balloon
   - J A golf ball

10. How are a lever and a ramp alike?
    - A Both are forces.
    - B Both have a fulcrum.
    - (C) Both are simple machines.
    - D Both have a slanted surface.

11. What happens to a ball when the force on it gets stronger?
    - (F) It moves faster.
    - G It stops moving.
    - H It begins to zigzag.
    - J It slows down.

Name _____

Chapter Test
Chapter 11

Write the answer on the line.

12. Why is it harder to ride a bicycle over grass than over a sidewalk?

   There is more friction on rough surfaces than smooth ones.
   _____
   _____

13. Why is it harder to push something up a short, steep ramp than up a long, less steep one?

   It takes more force to push something up the short, steep ramp.
   _____
   _____

14. Why is it harder to pull a sled up a hill than down a hill?

   Gravity pulls on the sled as you go uphill.
   _____
   _____

Name _____

Chapter Test
Chapter 11

15. Draw a picture of a lever.

16. Draw a picture of a ramp.

17. Why do people use levers and ramps?

Possible answer: To use less force when moving an object.

Forces and Magnets

Chapter Test — Chapter 12

Vocabulary

attract   compass   magnetic field
poles   repel

Use these words once for items 1 to 5.

1. Places where the pull of a magnet is strongest are called ____poles____.

2. A magnet will pull, or ____attract____, iron.

3. The area around a magnet where its force pulls is called a ____magnetic field____.

4. The same poles of two different magnets push away from each other or ____repel____.

5. You can figure out which direction you are facing with a ____compass____.

Name _____

Chapter Test
Chapter 12

Circle the letter of the best answer.

6. Which object will a magnet attract?
   A pencil
   (B) paper clip
   C shirt
   D milk carton

7. Which sentence is true?
   F All magnets are the same shape.
   G All magnets have the same pulling strength.
   H All magnets are the same size.
   (J) All magnets have two poles.

8. Which magnet has the strongest force?
   A A magnet that is far from an iron object.
   B A magnet that is far from a wooden object.
   (C) A magnet that is close to an iron object.
   D A magnet that is close to a wooden object.

9. What does a compass needle point to?
   (F) Earth's North Pole
   G space
   H Earth's South Pole
   J Sun

Name _____

Chapter Test — Chapter 12

Write the answer on the line.

10. Name two ways that you use magnets.

   Possible answers include to attach items to a refrigerator door, to lift items, and as parts of a game.

11. How is Earth like a giant magnet?

   Earth has a north pole, a south pole, and a magnetic field.

12. Name two objects that are attracted to a magnet.

   Possible answers include pins, nails, paper clips, or other objects made with iron.

Unit F · Watch It Move — Use with textbook pages F34–F53

Name _____

Chapter Test — Chapter 12

13. Draw a picture of two magnets that are attracted to each other. Label the north pole and south pole of each magnet.

14. Draw a picture of two magnets that repel each other. Label the north pole and south pole of each magnet.

Compare Smooth and Rough

Teacher Resources for Unit Performance Assessment — Chapter 11

Materials

- sheet of cardboard, eraser, bottle top, coin, rough stone

Objects can vary slightly from the list given. Try to provide students with items of various textures.

Scoring Rubric

5 — **5 points** Student correctly compares texture and motion of the objects and records all observations in the table. Student clearly explains his or her answers to the questions. Student infers texture of an object causes the amount of friction acting on it.

4 — **4 points** Student correctly compares texture and motion of the objects and records all observations in the table. Student clearly explains his or her answers to the questions. Student response to the inference is flawed but partially correct.

3 — **3 points** Student correctly compares texture and motion of the objects and records all observations in the table. Student response to the questions and the inference is flawed but partially correct.

2 — **2 points** Student comparisons of texture and motion of the objects contain one error. Answers to questions and inference are illogical or do not answer the questions.

1 — **1 point** Student comparisons of texture and motion of the objects contain two or more errors. Answers to questions and inference are illogical or do not answer the questions.

0 — **0 points** Student does not complete the assignment.

How Far?

Teacher Resources for Unit Performance Assessment
Chapter 12

Materials

- magnets of various size, paper clip, ruler

Try to provide students with magnets of various shapes as well as sizes. Some students may need assistance when drawing lines of a specific length.

Scoring Rubric

5   **5 points** Student correctly draws lines of specific lengths. Student records observations in the table. Student clearly explains his or her answers to the questions.

4   **4 points** Student correctly draws lines of specific lengths. Student records observations in the table. Answers to questions are flawed but partly correct.

3   **3 points** Student correctly draws two lines of specific lengths. Student records observations in the table. Answers to questions are illogical or do not answer the questions.

2   **2 points** Student correctly draws two lines of specific lengths. Student partially records observations in the table. Answers to questions are illogical or do not answer the questions.

1   **1 point** Student correctly draws one line of specified length. Student partially records observations in the table. Student does not answer the questions.

0   **0 points** Student does not complete the assignment.

Name _____

Unit Performance Assessment

Chapter 11

Compare Smooth and Rough

What to do

Compare Your teacher has given you four objects. Gently rub your fingers across them. Order the objects from smoothest to roughest.

Place the objects on one end of the cardboard. Slowly lift this end of the board. Compare the motion of the objects. Record the results in the table. Write 1 by the object that moved first. object that moved next. Write 3 by the object that moved third. Write 4 by the object that moved last.

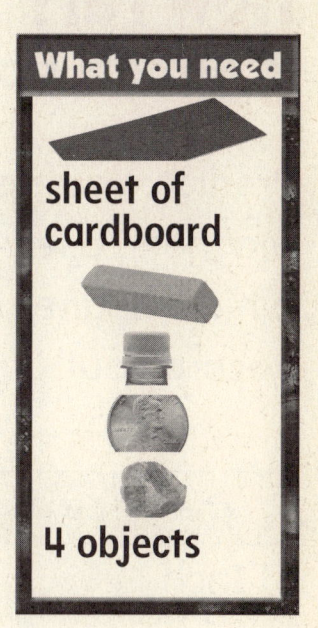

What you need

sheet of cardboard

4 objects

Repeat the test. Record those results.

Object	Test 1	Test 2

Analyze the Results

1. Order the items from fastest to slowest movement.

   Responses will likely be bottle cap, coin, stone, eraser.

2. How does your answer above compare with the objects ordered from smoothest to roughest?

   Students will likely note the same order exists.

Unit F · Watch It Move   Use with textbook pages F2–F33

Unit Performance Assessment
Chapter 12

Name _____

How Far?
What to do

What you need

magnets

ruler

Predict Your teacher has given you three magnets. Use your observations to predict which magnet has the strongest magnetic field. Record your prediction.

Distance	Magnet 1	Magnet 2	Magnet 3
1 Inch			
3 Inches			

Test your prediction. Draw a line 1 inch long on a piece of paper. Put a paper clip at one end of the line. Place a magnet at the other end of the line. Place the magnet so that one pole is pointing toward the clip. Record your observations in the table. Then test the other two magnets.

Draw a line 3 inches long on the paper. Repeat steps from above. Record your observations. Then test the other two magnets.

Analyze the Results

1. Which magnet has the strongest magnetic field?

   Answer should describe the largest magnet.

2. How does distance affect your results?

   The strength of the magnet will be weaker when farther away from the object.

72   Unit F · Watch It Move   Use with textbook pages F34–F53

Scoring Chart

DIRECTIONS: To convert a raw score into a percentage score, find the column that indicates the total number of items. Then find the row that matches the number of items that the student answered correctly. The intersection of the two rows gives the percent correct.

NUMBER CORRECT	\multicolumn{21}{c}{TOTAL NUMBER OF ITEMS}
	5	6	7	8	9	10	11	12	13	14	15	16	17	18	19	20	21	22	23	24	25
1	20	17	14	13	11	10	9	8	8	7	7	6	6	6	5	5	5	5	4	4	4
2	40	33	29	25	22	20	18	17	15	14	13	13	12	11	11	10	10	9	9	8	8
3	60	50	43	38	33	30	27	25	23	21	20	19	18	17	16	15	14	14	13	13	12
4	80	67	57	50	44	40	36	33	31	29	27	25	24	22	21	20	19	18	17	17	16
5	100	83	71	63	56	50	45	42	38	36	33	31	29	28	26	25	24	23	22	21	20
6		100	86	75	67	60	55	50	46	43	40	38	35	33	32	30	29	27	26	25	24
7			100	88	78	70	64	58	54	50	47	44	41	39	37	35	33	32	30	29	28
8				100	89	80	73	67	62	57	53	50	47	47	42	40	38	36	35	33	32
9					100	90	82	75	69	64	60	56	53	50	47	45	43	41	39	38	36
10						100	91	83	77	71	67	63	59	56	53	50	48	45	43	42	40
11							100	92	85	79	73	69	65	61	58	55	52	50	48	46	44
12								100	92	86	80	75	71	67	63	60	57	55	52	50	48
13									100	93	87	81	76	72	68	65	62	59	57	54	52
14										100	93	82	82	78	74	70	67	64	61	58	56
15											100	94	88	83	79	75	71	68	65	63	60
16												100	94	89	84	80	76	73	70	67	64
17													100	94	89	85	81	77	74	71	68
18														100	95	90	86	82	78	75	
19															100	95	90	86	83	79	76
20																100	95	91	87	83	80
21																	100	95	91	88	84
22																		100	96	92	88
23																			100	96	92
24																				100	96
25																					100

Units A–F   Use with textbook pages A2–F53

WITHDRAWN